Ocean Energy
White Paper

By Franklin C. Kostenko

Wednesday, April 30, 2014

Contents

1. Introduction

This paper aims to explore renewable energy devices and technology that harvest electrical energy from ocean currents and waves. This ocean born technology can greatly benefit private and industrial applications. For amplifying details and additional imagery please see the accompanying power point briefing, entitled Ocean Energy Technology.

Ocean currents and waves are an underutilized means of energy production today. For example, Maine has an energy density of 40 KW/m, which is enough wave energy to produce 1 megawatt of power with just 25 meters of coastline. Ocean currents in the Gulf Stream could provide 40 KW/m. Today less than 1% of the world's energy production comes from the ocean. If these sources were utilized, the US could possibly receive close to 40% of its energy from the seas. (2) These energy sources can also be used to provide energy for non-traditional power needs (e.g. buoy stations, transiting ships, houseboats, radio transmitter stations, military applications and aids to sea and air navigation).

Coastal States Department of Energy, our U.S. Sea Services and numerous major university's (Including University of Washington, MIT, Olin and MMA) provide opportunities for education and conduct research of sea power. With only two offshore ocean current harvesters in operation as of this writing, there is still an enormous need for more research in this field to design a more efficient sea energy system and to consider more duel use deployments is the future.

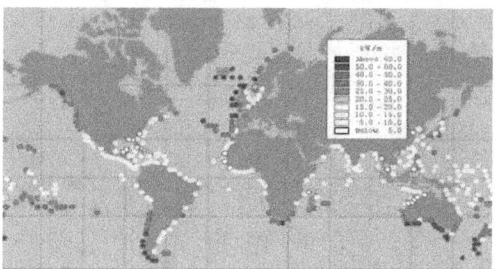

The chart above depicts the estimated energy potential globally.

2. Motive/Needs/Benefits

A crude picture of the available energy globally is above. Today offshore rigs require deep-sea air or surface travel. Such travels are expensive, time consuming and dangerous. Self-powered oil platforms could allow for remote monitoring while maintaining essential operation functions. Thus, travel becomes necessary only in the case of an emergency or at scheduled maintenance intervals. This could greatly reduce the cost associated with oil platforms. The field of electric power generation keeps gaining importance as more regions around the world obtain their electric power from renewable sources including wind and rivers.

The field of energy harvesting from water sources, other than dams, is relatively new. There are a few companies that have designed and deployed turbine-based systems to harvest water flow energy. (Wave and tidal) As of the writing of this paper, only one design attempts to harvest energy at an offshore location, and this particular design required large startup costs in order to fix the generator on the seafloor. The design of a cost effective, easy to deploy, water current generator will have a strong impact in the field. (11)

The energy present in the movement of ocean water around an oil platform is much larger than the energy required to maintain the essential functions of the oil platform. (3)

Competing harvesting technologies still must be analyzed based on the power requirements at hand. The current family of systems capable of fulfilling these requirements must still be specified. Detail design regarding said systems including the design of a mechanical system capable of turning fluid motion into mechanical motion, an electromechanical transducer which will convert mechanical energy into electrical energy, and an energy storage system which can store the excess energy and deliver it when the power source does not provide enough power.

Another creative use could be an effective "current flow" energy harvester. It could be adapted to power micro sensors attached to marine wildlife, enabling longer lifespan and greater range of these deployed research sensors. A novel method of capturing the flow may possibly be reversed to create a new type of water propulsion system. Global cruse lines and sea-going heavy transportation industries are looking at this creative concept closely. Scientists and engineers in Europe have been exploring ways to exploit the enormous amount of energy in ocean tides, currents, and waves since the 1960s. In the 1970s as a result of oil shortages research into ocean power technologies picked up significantly. Numerous patents were granted as many universities and firms began to develop prototypes of renewable energy systems to harvest power from the ocean. The majority of these systems soon proved unrealistic and unprofitable as the research progressed. (11)

Wave energy systems were often hindered by the shear size required of the system to produce reasonable amounts of energy. Meanwhile the development of marine current energy systems was halted by the cumbersome logistics of installing and maintaining systems in areas of high sea current flow, in addition to transmitting the power to the beach from remote blue water locations. Recently, advances in power transmission, energy conversion efficiencies, and advanced materials development, in combination with higher fuel and electricity prices have caused the reemergence of research in ocean power technologies. (2, 7)

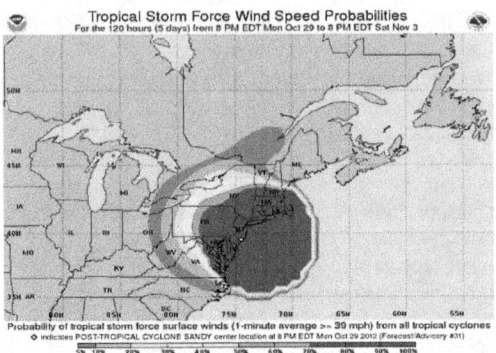

National Hurricane Center graphics on October 29, 2012

The recent major east coast storm, Hurricane Sandy has caused states to consider large seawalls and sea barriers to help break rouge and storm waves and control large storm surge effects. (Sandy brought 14-16 foot sea surge to the effected area). These projects would be enormous, yet if constructed and energy recovery was deployed into these coastal sea shields, this could help recover the massive cost of between $7 billion to an unheard of $200 billion dollars for a major metropolis projects being discussed such as off the coast of New Jersey and New York. Hurricane Sandy alone caused a reported $34.5 billion dollars in loss of property, so the duel use concept could be a very positive way ahead to help get such a massive project started and eventually deployed. (4)

3. Research Opportunities

Many of the ocean energy system ideas and different concepts presented in the accompanying power point presentation pertinent to extracting energy from the movements of the ocean water have existed for years. However, not until recently has it started to emerge as an enterprise with a future and find place in the worlds energy market. The aim of future research would be to improve the efficiency of existing technology and develop new technology that will offer greater utility and range for energy harvesting applications in the ocean. In contrast to other means of energy procurement, the processes and systems for extracting energy from the ocean are nascent. [6] While larger scale efforts have begun in Europe such as the Seaflow project conducted off the coast of the Britain by the UK and German governments, there exists to date only a modicum of industrial and government funding in the United States for ocean engineering research. [7]

Many of the present limitations, inefficiencies and problems subsumed in the processes of extracting energy from the ocean are a result of its seminal stage. Naturally, as more research is directed toward solving the current problems afflicting these technologies, the overall progress, efficiency, and lucrative nature of these endeavors will increase. Engineering research must join the body of technology and science that has coalesced around the problem of extracting clean energy from the environment. [7,11] While these solutions and benefits will be applicable to systems such as military applications and oil recovery platforms, they will also lend benefit to increasing the efficiency of other ocean energy systems for the mass populations.

The focus for the best way ahead is on usable techniques, methods and systems. Academic and engineering research should be guyed by the applications and promise of providing power to a myriad of scientific and industrial applications. The goals of innovation and improvement on current systems will be shored by an acute awareness of how this work contributes to the larger and critical issues of yield, efficiency, reliability, required maintenance, and the associated costs. Understand that in today's world a sound engineering solution is also a fiscally viable one that assumes feasible manufacturing, installation and reasonable maintenance costs. (12) It is also worth address reliability issues centered about common ocean structure failures such as brittleness and fatigue, scour, dynamic and static loading, erosion, fretting, corrosion and bio-fouling and of course, plain and simple loss of equipment during strong seas and storms.

Critical properties of ocean energy sources like waves and currents vary over time. By focusing on the development of systems that would sense and adapt in real-time to the inherent fluctuation of an energy source by continually tuning mechanical and electrical properties to ones environment to optimize efficiency— one can increase the overall efficiency of such an ocean system and thus increase energy capture. (5)

Powering no longer yielding oil platforms by means of renewable energy from ocean sources could provide an excellent resource for marine scientists and engineers. Boat or Sealift time is perhaps the mostly costly component of ocean research expeditions. In cases where scientists or engineers need a ship to travel to and stay at one spot for an extended period of time it may save a lot of money to instead use the ship only for traveling to and from an oil platform. Then the scientists or engineers could do their research while living on the platform. For example, in researching ocean power technologies a group of engineers could install several different technologies on and around the platform and then assess and maintain the devices as needed as they live on site for a period of time. (8)

Ocean power technologies developed for deployment onto obsolete oil platforms could also be implemented on a working platform, decreasing the platform's energy costs and environmental effects. Technologies developed for this application will be valuable prototypes for larger-scale ocean power generation projects. In a future where renewable energy may be the primary method for addressing the world's energy needs, developing ocean power technologies now is an important and necessary step for our future.

The focus of capturing energy from the motion of the ocean segues to areas of research that demonstrate the scalability of the technology one intends to develop. Opposite the macro, megawatts scale of harvesting energy is the venture of developing devices to capture energy in the micro-scale regime of watts and kilowatts for scientific instrumentation. (1, 5)

Ocean and Naval force data is collected by a myriad of remote scientific instruments such as monitors on docks, buoys and moorings. The utility of the navigational buoy is most significant in the extreme and distant locations where travel and first hand monitoring is expensive and dangerous, or even in a non-permissive location where it is not possible at all. Not surprisingly, many of these locations are home to active or high-energy sea conditions. Currently these buoys and moorings are powered by easily broken and poorly maintained solar panels and battery systems. (5) It seems feasible that technology aimed at deriving energy from the ocean could be implemented to reduce maintenance costs and increase the available power to these buoys and remote dock sites.

Another idea I'd like to put out, is the concept of harnessing the energy of an animal as it swims through the water—in essence turning the animal into a power plant— to power monitoring tags and other devices currently used by the military or scientist to study pelagic animals. Provided the animal is sufficiently large and provides a critical minimum amount of convertible mechanical energy, from either flow around the body or body movement, this seems plausible. Such a power supply could replace the batteries found in current seabed acoustic systems; overhead satellites or pop-off archival transponder tags relied upon to study ocean traffic and aquatic life absent from overhead monitors. (8) Used in concert with existing methods, such a tool could expand the range and depth of intelligence by gathering more and new types of data for longer periods of time. (7)

There are several animals large and powerful enough for this method to work quite well— e.g., certain species of shark and tuna known as obligatory ram ventilators must always be swimming at some minimum speed so to pass a sufficient amount of water through their gills. However, it seems reasonable that with specialization this method could be used to provide power for monitoring marine life of a variety of shapes and sizes. Such developments could offer additional advances in the field of remote ocean sensors. They could also have military applications that will not be discussed further in this form. (1, 5)

4. Ocean Power Specifications

4.1 Energy Conversion

The basics. The figure below is a system diagram of how energy can generally be extracted from an ocean environment. Power in the form of a fluid (kinetic, potential, pressure, etc.) energy comes from the wind, waves, currents, etc. Transferring fluid energy into a rotary or oscillatory motion through hydrodynamic forcing then creates mechanical energy. (See accompanying power point brief, Attachment #1, for visual examples of numerous options in use today) The mechanical energy is then converted into electrical energy by means of an electro-mechanical device (e.g. a generator). The raw electrical energy is then conditioned, converting it into useable clean electricity. There are also devices that can directly convert fluid energy into electrical energy through peizo-electric effects or electro-magnetism, thus removing the mechanical energy transfer step in figure below.

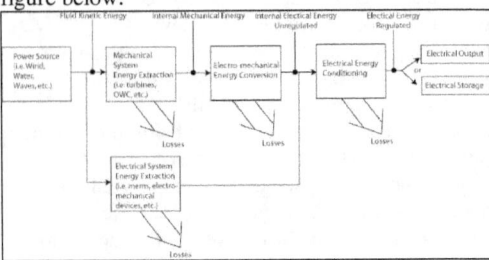

Energy transfer diagram.

This figure also illustrates system energy losses at each conversion step. Minimizing losses is important to the development of viable energy technologies. Wavegen©, a company producing power from wave energy off the coast of Scotland, underestimated their system losses by as much as 89%. During the design process it is important to account for losses properly in order to calculate a realistic overall system efficiency for comparison to other energy technologies. (1,2)

4.2 Ocean Currents

The power in a flowing fluid (i.e. wind, water currents, etc.) is determined by the flux of kinetic energy passing through an area (A). That is the mass flow rate (ρUA) multiplied by the wind kinetic energy per unit mass ($U^2/2$) see equation (A).

$$P = \frac{1}{2} \rho A U^3$$

(A)

Ocean currents are usually classified as wind driven (typically surface currents) or thermohaline (typically underwater currents). Ocean currents are often layered; deep-water currents can travel in a completely different direction and at a different speed than the surface currents. The figure below shows currents for the world's oceans. To understand how much power ocean currents can contain, assume an average current of 1.5 m/s found in the gulf stream, in order to get the same energy in wind the speed would need to be about 14 m/s.

4.3 Waves

Wave power per unit width is calculated by integrating the pressure under the wave in equations B. The pressure difference (p_1-p_2) and velocity (u) are defined here.

$$P = \frac{\rho g^2 a^2}{4\omega}$$

(B)

The average wave power increases as the amplitude (a) squared and the inverse of the frequency (ω).

The figure below shows an average wave power profile for the world's coastlines. As pointed out in the opening of this paper, Maine has an energy density of 40 KW/m, which is enough wave energy in to produce 1 Megawatt of power with just 25 meters of coastline. For a typical wave energy efficiency of 10% the same 1 Megawatt would require about 250 meters of coastline.

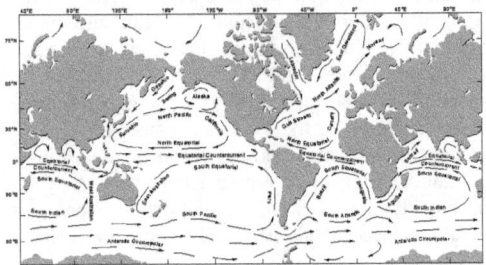

5. Past and Present Technologies

5.1 Ocean Current Technology

Tidal current power technologies generally rely on the use of marine turbines to capture power from water flow. The accompanying PowerPoint slide deck contains a visual displace of a number of examples in used. The La Rance Tidal Power Plant, in Brittany, France has been producing power since the mid 1960's. Maintenance costs have plagued the system since the beginning, which is probably why ocean technology has not been exploited much throughout the world as of yet. Several companies such as Blue Energy (Canada), Tidal Hydraulic Generators Ltd (UK), and Woodshed Technologies (Australia) are currently working on creating low maintenance advanced turbine systems, which are easier to service, in an effort to make tidal power a more promising renewable energy option. (2, 13)

Image and graph below of the La Rance Tidal Power Plant, in Brittany, France

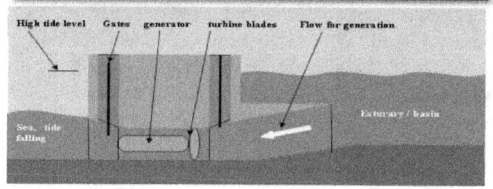

Marine Current Power (UK), Verdant Power (US), and SMD Hydro vision (UK) are developing special turbines/propellers optimized for low-speed, open-water currents. Another company, The Engineering Business Ltd (UK) has taken a different approach and designed a device called the 'Stingray,' consisting of a hydroplane that is connected to a fixed structure through a hydraulic joint. The hydroplane oscillates as the water flows past it powering a generator. The company recently stopped research on this project due to its projected unprofitability. The remoteness of feasible marine current sites leads to problems in efficient power transmission and machinery maintenance. (2,13)

5.2 Wave Technology

One way to generate power from waves is through use of an oscillating water column (OWC). Waves encountered by an OWC cause the water level to rise and fall within the main chamber forcing air through turbines, generating power. Both fixed and floating OWC devices have been developed. Daedalus (Greece) and Wavegen (UK) have both built and implemented fixed OWCs. Wavegen, probably the leading OWC developer, is now working on a large-scale project to install an enormous fixed OWC within a coastal cliff on Faroes Island. Other companies such as ORECon (UK), in addition to the Japanese Agency for Marine Earth Science and Technology (JAMSTEC) have developed floating OWCs. (6)

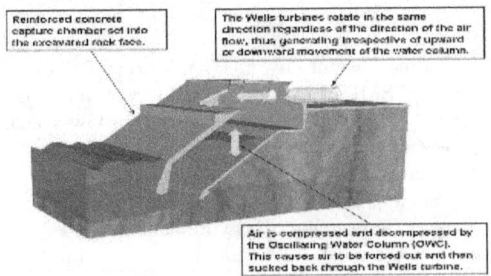

Wave energy can be extracted by a series of hydraulically connected floating rafts. As the rafts move relative to one another hydraulic fluid is pumped back and forth and this forced motion can be used to generate power. This method of ocean power extraction was first envisioned by Sir Richard Cockerell, an English inventor in about 1972 (12). However, significant advances in materials and hydraulic control systems have only recently made this technology a realistic prospect. Ocean Power Delivery Ltd. has spent the last five years developing Pelamis, a 150m long device. The three joints connecting its four discrete sections derive power from both the roll and pitch of the device as it follows the waves. The Pelamis has been thoroughly tested in the North Sea and recently the company earned its first contract, from the Portuguese government. Allianz is also producing the Sea Snake design out of the United Kingdom. (8)

Pictured above is the Sea Snake, AKA: Pelamis, by Allianz

A third method for wave power extraction is a linear generator. These devices consist of a fixed internal unit and a floating external hood. The fixed unit is tied to the seafloor and contains a large permanent magnet. The hood, surrounded by a wire coil, is free to heave. The relative motion of the coil and magnet produces an electrical current. Tuning the system to the average wave frequency produces resonance. Two companies have been working on this technology: Wave Power Technologies (US) and Archimedes WaveSwing (Netherlands). (1, 13)

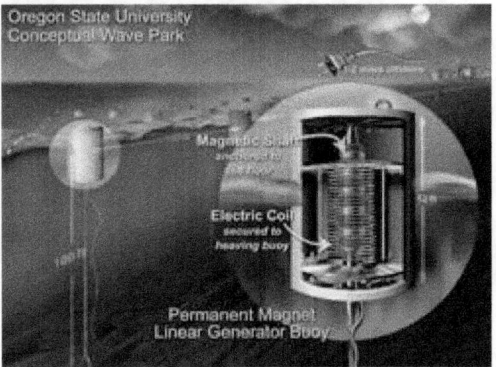

Pictured above is a rendition of a Wave Farm made up of Permanent Magnet Linear Generator Buoys.

6. Exploration Means

This section could also be called "the way ahead" or where I would start to get this technology off the ground. Extensive maritime equipment can be found and if required from Woods Hole

Oceanographic Institute (WHOI). This is the largest independent oceanographic institute in the world today. It has three large research vessels allowing access to the open seas, full testing facilities including a twenty-thousand psi pressure test chamber, complete machine, welding and metal fabrication shops and over 130 full-time scientists and engineers. It is located directly on the ocean providing the unique ability to easily essay objects in a marine environment. (6, 10)

The facilities at the Maine Maritime Academy (MMA) provide a unique feature of the MMA facility is the presence of a significant tidal current at the deep-water pier at the Academy. This site could be used as a convenient current energy extraction test site. Dockside facilities include typical heavy equipment found at a typical small vessel boatyard, including welding and machine shop facilities. MMA has over 20 full time engineering faculty and 400 undergraduate students in the engineering programs. Various small vessels including a small research vessel, a large barge and a fully operational tugboat are owned and operated by the Academy. (13)

In particular the United States Navy will look harder at the use of Ocean Energy. It is in the sea services roots to use the power of the sea for her mission, back from the start when in Philadelphia, Oct. 13, 1775, they developed the navy's first two armed sailing vessels. The goal of these sailing ships was to intercept transports carrying munitions and supplies to the British army in America. In fact the US Navy is an energy leader, they operate and maintain the most formidable fleet of nuclear-powered submarines and aircraft carriers on the planet. They apply nuclear energy and fundamentals of engineering in ways that not only help to defend our national security but also serve to better our world. (5) I'm confident the future of Ocean Energy will be dead ahead on the Navy's radar. Congress has been refusing to let the US Navy continues tests with bio fuels, most believe this is motivated by politics, as the Nuclear Navy has always been very forward leaning with regards to energy and use of ocean and other natural resources since it's birth.

7. Conclusion

I have enjoyed reading several books and looking over many reports on this subject over the past few months, from my very pedestrian position on the matter, I'd say Ocean Energy does have a bright future. I feel strongly at this time we should focus on the duel use systems foremost. In other words, double up use where major ocean engineering projects are already being scoped and deployed. I focused on the two main sectors of waves and tidal and current technology. However under the title of Ocean Energy we must not forget the salinity power and thermal energy options. I will briefly touch on these two more complicated areas while closing.

The mixing of freshwater and seawater releases energy. The challenge is to capture and utilize this energy, since the energy released from the occurring mixing only gives a very small increase in the local temperature of the water. (14) During the last few decades at least two concepts for converting this energy into electricity instead of heat has been identified. These are Reversed Electro dialysis and Pressure Retarded Osmosis. With the use of one or both these technology one might be able to utilize the enormous potential of a new, renewable energy source. On global basis this potential represents the production of more than 1600 TWh of electricity per year.

Ever since Georges Claude conducted his pioneering work on Ocean Thermal Energy Conversion (OTEC) nearly 80 years ago, engineers have dreamt of tapping this enormous renewable resource. The ocean thermal gradient could be used not only to produce electricity, but also in derivative technologies like desalination, cooling and aquaculture. With ocean thermal energy the overall sustainable size of this resource is limited by the rate of formation of deep cold seawater. Orders of magnitude between 3 and 10 TW appear likely, a range approximately ranging from twice today's overall electricity consumption to about half of today's primary energy needs. (14)

Multiple references (below) have stated they believe the United States will be using about 15% ocean generated energy nationally by the year 2030.

The ideas of placing ocean energy system at remote deep sea locations seems to make logical sense if the site is also required for an additional objective, such as communications and/or navigational requirements, remote ocean floor monitoring, fish and other sea life migration, and acoustic intelligence operations.

The permanently placed shore based power systems would need to be built so robustly to withstand the constant sea wave ponding, that currently this renewable may be too costly for common use to make financial sense. However costs via multi use projects is making this option more viable. For military and humanitarian short term deployment of these systems, the use is feasible and in use now.

The same is true of the US Governments vast array of weather and ship tracking sensors on docks and navigational buoys globally, these heavy duty ocean energy systems for targeted use is not only a good idea, it is a must as the current use of solar and batteries is a failing proposition due to these systems being constantly damaged and degraded by wild life and storms. I look forward to monitoring this exciting field over the coming years, and watch as it expands into more common use.

Research References

1   Wavegen©, 2002,
    http://www.wavegen.co.uk/research_papers.htm

Crown Publishing, *Islay Limpet Project Monitoring Final Report*, p 34.

2   Energy for keeps 3d Edition, 2006, Energy Education Group, *Nemzer, Page, Carter*

3   Energy Insight, http://www.energyinsights.net/cgi-script/csArticles/articles/000019/001983.htm

4   Business Week, Nov 5, 2012 issue; http://www.businessweek.com/news/2012-11-05/taxpayer-storm-shield-protects-casinos-while-poor-take-on-water

5   Office of Naval Intelligence (ONI) Open source content ref. 1 JUN 2011 Energy Options. http://www.oni.navy.mil Navy Energy

6   MDA Home page, http://www.marinedevelopmentinc.com/ocean_energy.htm 05 OCT 2012

7   University of Washington, white paper on Ocean Energy. 1 JAN 2010

8   National Maritime Intelligence Institute, Suitland Park MD

9   University of St. Marry, St. Louis MS. Neal A. Allen, White Pater of 17 Apr 2010

10 Alexandra Techet, Professor of Ocean and Mechanical Engineering at the Massachusetts Institute of Technology in Cambridge, MA,

11 Franklin W. Olin School of Engineering in Needham, MA, Jose Oscar Mur-Miranda, Visiting Assistant Professor of Electrical and Computer Engineering. Sea Renewables White Paper 02 Sep 2012.

12 Arkwright by Sir Richard. http://www.infoplease.com/encyclopedia/people/arkwright-sir-richard.html

13 OCS and Alternative Energy and Use
Progammatic EIS, Info Center.
http://ocsenergy.anl.gov/guide/wave/index.cfm

14 Ocean Energy Systems, http://www.ocean-
energy-systems.org

www.ingramcontent.com/pod-product-compliance
Lightning Source LLC
Chambersburg PA
CBHW070735180526
45167CB00004B/1763